On 16 August 1911 Lt. H. R. P. Reynolds arrived at Port Meadow, Oxford in Bristol Boxkite 'F7' after a flight from Larkhill on Salisbury Plain as part of a cross-country exercise. Three days later he departed for Thetford, but was forced down at Launton, near Bicester, giving the local populace what was possibly their first close up view of an aircraft. Reynolds took off again later in the day but crashed near Bletchley. Port Meadow became a large training airfield during the First World War, and was joined a few months before the cessation of hostilities by Bicester, Upper Heyford, Weston-on-the-Green and Witney. The site selected for a training station near Bicester was at Caversfield, a little under two miles north of the town, and construction commenced in 1916.

Commensurate with the airfield construction was a railway spur from the Oxford to Cambridge line near Launton station, about a mile east of the centre of Bicester, and this was initially used to bring in most of the building materials, and remained in use until the late 1930s. The site of the new airfield straddled the Bicester to Buckingham road, the flying field and technical site being to the east and the domestic site to the west along Skimmingdish Lane. As on many airfields constructed at this time, labour came from a number of sources. Under the charge of the Royal Engineers, the work force included Canadian sappers, Portuguese and Chinese labour groups and German prisoners of war, while in 1917 the US Army Corps of Engineers installed an electric power station, the first in the district.

Bicester airfield opened at the end of 1917 for use as a training Station in Southern Army Command. The site, with a landing area of 3450ft. by 3000ft. (1050m. by 915m.), covered a total of 180 acres (72.9 ha.), thirty acres of which were occupied by buildings. The original hangars, erected in 1917, were the canvas 'Bessonneau' type but during 1918 three pairs of general aeroplane sheds and a single aeroplane repair shed were constructed alongside the Buckingham road.

118 Sqn., destined to be a night bomber unit, moved in from Catterick with Bristol F2b Fighters on 7 August 1918. On 1 October, 44 Training Depot Station (21st. Wing) moved in from Port Meadow with the role of fighter and reconnaissance training. 118 Sqn. was destined to be equipped with Handley Page 0/400s, but due to cessation of hostilities, it was disbanded in November before becoming operational. On 12 February 1919, 2 Sqn., a reconnaissance unit, moved in from Genech, France, with Armstrong Whitworth FK8s. Its stay was brief as it made the short move to Weston-on-the-Green in the September, making way for 5 Sqn., which arrived from Hangelaar, Germany on 8 September with Bristol Fighters. 44 TDS was redesignated 44 Training School in August 1919 but was short lived with this title, being disbanded in December. On 20 January 1920, 5 Sqn. also disbanded.

Bicester closes its gates

On return to peace Bicester also became a clearing house for repatriated British prisoners of war and remained open until official closure on 17 March 1920, after which the wartime buildings were soon demolished. During 1925 Bicester was earmarked as a three squadron bomber airfield with six permanent hangars, but during reconstruction of the airfield in 1926, defence cuts had reduced the number of aircraft in a squadron from 18 to 12 and so only two hangars were erected. As building work was well under way, this left the Station in the odd situation of having technical buildings to support three squadrons and hangar accommodation for only two, although there was space for additional hangars should they be required at a later date.

The planned new Station required a greater area of land and fields known as 'Barnfield' and 'Skimmingdish' were acquired to the west of the airfield, to extend the boundaries of the domestic site, and Hungerhill Farm was acquired to allow extension of the airfield to the north. An additional strip of land was also acquired for extension of the

No.2 hangar at Bicester, an A-type building, with a Horsley of 100 Sqn. outside and the railway spur in the foreground, seen in about 1929.

railway line.

During the winter of 1926/27, over the Buckinghamshire border at the RAF Apprentice School at Halton, members of the Halton Aero Club was busy constructing their own aircraft, the Halton HAC.1 Mayfly. Possibly due to Bicester being a more suitable airfield for winter flying, the Mayfly (G-EBOO) was transported by road and reassembled at Bicester, where the maiden flight was made on 31 January 1927.

Bicester reopens in the bomber role

By 1927, there was a maximum take-off run of 4170ft. (1271m.) and the field was identified by a 150-foot (45.7m.) diameter chalk landing circle. By now the railway spur terminated at the main stores and a concrete compass-swinging platform had been built. The lack of flying units on the Station was remedied on 1 December 1927, when an advance party of 100 (Bomber) Sqn. arrived from Spittlegate, during one of the severest winters on record. 100 Sqn. was a Day Bomber unit equipped with large Hawker Horsley two-seat biplanes, and formed part of the 'Wessex Bombing Area' of the Air Defence of Great Britain scheme. The arrival of the main party was delayed when the water systems in the barrack blocks froze solid but they eventually made it on 10 January 1928. The CO, Sqn. Ldr. L. T.

Seen being refuelled outside a Bicester hangar in 1928 is a Horsley of 100 Sqn. (possibly J7998) Note the fuel pipes suspended from overhead cables [D. S. W. Blee collection]

BICESTER

N. Gould MC, also assumed command of the Station. From mid May until mid June the squadron moved as a unit to Weston Zoyland in Somerset for annual practice camp.

Horsley J8005 was lost in a mid-air collision over Upavon on 12 June 1930, when it collided with 3 Sqn. Bulldog J9574 during a practice for an air display. 100 Sqn. departed for Donibristle in November and two days later 33 Sqn., which had been the first unit to receive Harts the previous February, moved in from Eastchurch with these new mounts. On 1 October, Central Area was formed at Andover to control the bomber squadrons at Abingdon, Bicester, Bircham Newton, Filton, Hucknall and Upper Heyford and on the same day the Wessex Bombing Area took on the simpler title of Wessex Area. On 20 October 1931, 18 Sqn. had reformed at Upper Heyford and on 9 November 33 Sqn. gave up four Harts to form the nucleus of this new unit. In June 1933, 33 Sqn. Harts took part in displays at Andover and Hendon and remained based until

November 1934 when they moved to nearby Upper Heyford, and a little over a year later to the Middle East.

Their place was soon taken by 101 Sqn., which arrived from Andover in early December with Boulton Paul Sidestrand medium bombers. 101 had the distinction of being the only squadron so equipped at this time, and although ungainly in appearance and rather heavy on the controls, the Sidestrand was popular with aircrew. Classed as a medium bomber, when carrying a heavy bomb load it proved to have the speed and manoeuvrability of a single engined light bomber. It was also a superb bombing platform, taking all RAF bombing competition records. Only eighteen Sidestrands had been built and it was to be a redesign which would work alongside it on 101 Sqn., and eventually replace it. The high speed of the Sidestrand, exposing the nose gunner to a high chill factor, had caused problems when training guns on a target and changing ammunition drums with frozen fingers. Christened the Overstrand, the redesign was similar in appearance to its forerunner

One of 101 Sqn's. Overstrands, K4561 [101:U] with an invisible pilot and a prominent gunner! The photo illustrates the higher standard of protection than that on the earlier Sidestrand.

5

Seen in this photo of one aircraft taken from another in formation is Sidestrand Mk.III (originally Mk.II) J9178 [A] of 101 Sqn., with which it served between March 1929 and January 1936. K1994 (on left) was a similar aircraft, which carried code letter D [D. S. W. Blee collection]

but had been re-engined and had a revolving turret in the nose and a canopy over the cockpit, allowing a few creature comforts for the crew.

Initially Overstrands were converted from their forerunners but eventually an order was placed for new-build aircraft and 101 Sqn. was increased in size from two to three Flights. The new 'C' Flight received the Overstrands, the Sidestrands of 'A' and 'B' Flights being replaced as new aircraft were delivered. The first Overstrand conversion, J9185, was delivered to Bicester on 24 January 1935 and the new build aircraft arrived between October 1935 and July 1936. Overstrands had the distinction of being the last RAF biplane heavy bombers to enter service and the first aircraft in the world to have a power operated gun turret. On 6 July 101 Sqn. took part in the King's Jubilee Air Review at Mildenhall, and when inspecting the squadron His Majesty climbed aboard Overstrand J9185 and sampled the new turret. The squadron won the Armament

Officers' Trophy at the annual practice camp in 1935 for the second year and also the Sassoon Photographic Trophy (and again in 1936). On 9 September J9185 was lost when it crashed on the North Coates range. In November, personnel from 101 Sqn. 'C' Flight formed the nucleus for the reformation of 48 Sqn. and a few weeks later, still without aircraft, moved to Manston.

On 1 May 1936, Central Area became No 1 (Bomber) Group and in December a Station Headquarters was formed within Bomber Command in No 1 (Bomber) Group, 101 Sqn. being the only unit on charge. In September 1936 Overstrand J9179 nosed over at Bicester and was written off, and in November K4562 ended up in a similar attitude after a brake seizure. K4556 did likewise after force landing in boggy ground at Bicester.

33 Sqn. harts carried the squadron number in an unusual position above the fuselage roundel, as seen in this picture of K2443 and two others. In June 1935 K2443 became GI airframe 651M.

BICESTER

The Blenheim makes its debut

January 1937 saw the reformation of 144 Sqn. as a day bomber squadron from 'C' Flight of 101 Sqn., temporally equipped with four of the former's Overstrands but sadly losing K4564 on the same day when it crashed in fog in Buckinghamshire with the loss of three crew. By the end of the month it had changed these temporary mounts to Ansons and on 9 February it left Bicester for the newly opened airfield of Hemswell in Lincolnshire, where it initially received Audax, and later in the year, Blenheims. 15 March saw the emergence of yet another squadron from 101, when 90 Sqn. was reformed from 'B' flight. It was to be equipped with Blenheim Mk.Is, but due to production delays it initially received Hinds K6738-6750, delivered direct from the production line for use until the first Blenheims began to arrive in mid-May, when it became the second squadron to be so equipped. On 3 May, Sir Phillip Sassoon, Under Secretary of State for Air, visited the station. The first Blenheim incident was on 20th May, when K7050 of 90 Sqn. suffered an undercarriage retraction on the ground. The first fatal accident happened on 30 June when Blenheim dual trainer K7053 dived out of cloud over Shropshire during a navigational exercise and failed to recover.

Conversion to the Blenheim was not without problems. For pilots experienced on more simple types, the retractable undercarriage and variable pitch airscrews accounted for most of the difficulties. Many pilots ended up red-faced, sitting on the airfield surrounded by the emergency services after retracting the undercarriage instead of the flaps, as the levers for these operations were located side by side.

On 5 July, as part of the RAF Expansion plans, the construction of two Type C hangars (drawings 1581/35 and 2392/37), in front of the existing two commenced. In August, 217 Sqn., a general reconnaissance unit equipped with Ansons, arrived for a month's stay before returning to its Tangmere base. Throughout the summer Bicester had the appearance of a building site. As well as the new hangars, a bomb dump at the east of the airfield, an additional barrack block, additional officers and airmen's married quarters, a fire station and a watch office and tower (drawing 1959/34) were all taking shape. At the end of October a floodlight was taken on charge for night flying, a technique only then coming into prominence.

90 Sqn. Blenheim Mk.I K7052 stalled and force landed when on fuel consumption trials on 17 December. During the final Hendon air display, a 101 Sqn. Overstrand was air-to-air refuelled by a

Blenheim Mk.I K7052 of 90 Sqn. came to grief on 17 December 1937 between Middleton Stoney and Weston-on-the-Green while on fuel consumption trials. [D. S. W. Blee collection]

The unusual lines of a Battle Trainer of 12 Sqn., unfortunately without a clear serial number.

Vickers Virginia tanker, this providing a foretaste of 101's role today.

101 Sqn. crews had to look on at 90 Sqn. with envy for more than a year, while they soldiered on alongside them with their aging biplanes. At last, in June 1938, it was their turn to modernise, when the first Blenheims were delivered to the squadron, and re-equipment was completed by the end of August. At this time the Station Flight had a Magister on charge.

The September 1938 Munich crisis caused more than a little panic, during which the Station was brought to a high state of readiness. 144 tons of bombs were delivered by rail and service transport until a war state scale been achieved. All Bicester's Blenheims were fitted with dummy perspex turrets, a situation that was soon rectified by civilian working parties fitting new powered turrets on site. On 29 September the Station came under the control of 2 (Bomber) Group. The large white under-wing aircraft serial numbers were deleted at

this time and the complete undersurfaces of all aircraft painted matt black. In mid-December both 90 and 101 Sqns. carried out tactical exercises with Fighter Command, but due to inclement conditions the airfield was not fit for flying from the 23rd to the 28th of the month.

In March 1939, 90 Sqn. became the first RAF squadron to re-equip with the long-nose Blenheim Mk.IV, which was a great improvement over the Mk.I, having improved accommodation for the navigator. During May 90 and 101 Sqns. were transferred to West Raynham and were replaced by 12 and 142 Sqns., which moved in from Andover with Fairey Battles on 9 May. After a few days of dive bombing practice, on 20 May 12 Sqn's 'A' Flight demonstrated six Battles at the Empire Air Day displays at Upper Heyford and Ansty and 'B' Flight, also with six aircraft, at Halton and Henlow. 142 Sqn. had departed to the Temporary Armament Training Camp at Leuchars, Scotland on 13 May for a 21-day stay on its annual camp. Throughout the summer

Blenheim Mk.I K7092 [90:K] of 90 Sqn. at Bicester, with which it served from October 1937 to April 1939.

both squadrons were engaged in training sorties, including Regional Exercises which involved gas spraying with the Northern, Scottish and Western Army Commands. 142 Sqn. 'A' Flight was detached to Montrose for eleven days in July and in August the squadron operated out of Weston-on-the-Green for a week for Air Defence Exercises.

The outbreak of war at RAF Bicester

76 (Bomber) Wing reformed on 24 August from SHQ Bicester and on 1 September, 12 and 142 Sqns. became the aircraft component of 76 Wing of the Advanced Air Striking Force (AASF) and were ordered to prepare for mobilisation. On the same day a number of civilian transport aircraft arrived at Bicester, including HP.42s and Ensigns, and by nightfall they had been camouflaged by Station personnel. Next day at 14.15 a formation of sixteen Battles and one Magister of 12 Sqn. departed for Berry-au-Bac, France, where the ground crew advance-party of one sergeant and sixteen airmen had already arrived in the civilian transports. 142 Sqn. had similarly followed 12 Sqn. to the same location. At 11.00 on 3 September 1939 war was declared with Germany and the Battles taxied out and were bombed up, this time not for a practice. The AASF comprised 71, 72, 74, 75 and 76 Wings and all MT transport departed for France via Avonmouth on 11 September. The remaining squadron personnel left from Bicester railway station on 16/17

September for Southampton, once more leaving RAF Bicester almost silent, although an advance party from Bassingbourn had by this time arrived to take over the running of the airfield.

To relieve active squadrons of their training tasks, in 1939 it was decided to set up a number of Group Pools, withdrawing some squadrons from their operational roles and designating them as training units. 104 and 108 Sqns. from Bassingbourn, took up residence at Bicester in mid-September, both equipped with Blenheim Is and IVs and Ansons. Both were Group Training Squadrons tasked with training crews for operational squadrons, Bicester becoming 2 Group Pool, under the control of 6 Group, to supply crews for its Norfolk-based front line Blenheim squadrons. Its remit was to convert pilots to the Blenheim and to train pilots, observers and air gunners to become operational crews. A Station Defence Scheme came into being in October and the Station, which had been camouflaged by this time, received detachments from the Oxford and Buckinghamshire Light Infantry and the Royal Artillery. On 19 October the first detachment of WAAFs were posted in and were billeted at nearby Brashfield House and in Officers married quarters.

Halifax sojourn

The prototype Handley Page Halifax four engine bomber, L7244, had been completed at the

Prototype Halifax L7244 at Bicester in October 1939 during early trials. [D. S. W. Blee collection]

Cricklewood factory as war began. Both the Air Ministry and the company were in agreement that the company's Radlett airfield was too small for the first flight of such a large aircraft, as it offered a take-off run of only about 2250 feet (685m.), which left no margin for error if any problems were encountered. The nearest non-operational airfield at this time was Bicester, with no based aircraft but offering large hangars and all the facilities for re-erection and flight testing. L7244 was duly dismantled and transferred to Bicester by road and discreetly re-erected by a Handley Page team in the hangar closest to the Buckingham road which was used by 108 Sqn. Cordes, the chief pilot, made frequent visits in a Magister to follow the progress, being careful to make his visits seem a low-key affair for security reasons.

When the day came for taxiing trials, the results were not promising after only one run. The Lockheed hydraulic brakes proved to be slow-acting and Cordes refused to make the maiden flight until they had been replaced with Dunlop pneumatic units. The cat being firmly set among the pigeons, Handley Page himself was soon on the road to Bicester in his Rolls Royce, hotly pursued by C. D. Holland of the design team. Holland managed to overtake and arrive at Bicester ahead of his boss and make contact with the Dunlop representative, who in turn was able to make contact with his Coventry base before the irate HP arrived. Installation of the Dunlop pneumatic system took three days and three nights to complete, although by the time of the first flight, the air compressor had not been installed and air bottles had to be carried in the fuselage.

In the hands of Cordes, the Halifax lifted from Bicester's grass field on 25 October and made a successful first flight with the undercarriage locked down. In spite of all the secrecy surrounding its stay, the first flight was witnessed by a crowd of locals, much to the surprise of all concerned. All the preliminary handling tests were carried out at Bicester, and on one occasion an elevator fractured along a spar line, but as the centre of gravity was near the mid-range Cordes was able to make a safe landing. Bicester having served its purpose, L7244 departed for Boscombe Down for further trials. On completion of production in 1946 6,176 Halifaxes had been built.

The Group Pool had a successful first couple of months, completing 1344 day and 117 night flying hours in October and 1254 day and 74 night in November. Two 108 Sqn. Ansons collided in mid-air on 31 October and a Blenheim of the same squadron crashed at Bicester on 31 November. A 'K' site dummy airfield was set up at Grendon Underwood, which is south east of Bicester in Buckinghamshire, with full size dummy aircraft, while Bicester's flashing identification beacon was set up in a lay-by on the adjacent Aylesbury road.

The winter of 1939/40 was a very hard one, and the first victim was 108 Sqn. Blenheim Mk.I L4885, which returned to the field in poor weather conditions, landed with the flaps up and overshot into the railway embankment, luckily without injury. Weather permitting, Weston-on-the-Green was now used for bombing practice. By the end of December, after terrible weather conditions, the field became flooded and all flying training was carried out at Kidlington.

Trials and tribulations of 1940

The weather was no better early in 1940, due to freezing conditions, 22 degrees of frost being recorded at Bicester on 20 January. Flooding caused the airfield to become badly churned up and a subsequent freeze set all the ruts like stone, resulting in damage to the undercarriages of several taxiing Blenheims. As if conditions were not bad enough, heavy snow on 26 January put an end to flying on the Station until the end of the month. During February the airfield was unfit for flying for a total of nineteen days and total flying for the month amounted to only 195 hours.

In the late 1930s Blenheims had been delivered in small numbers to the air forces of Finland, Turkey and Yugoslavia. Finland received a batch of eighteen in 1937 and the following April was granted a licence to produce its own aircraft. Unfortunately, none were completed before the start of the Russo-Finnish 'Winter War' in November 1939 and so the British Government agreed to supply additional aircraft from RAF stocks. Following conversion to the Blenheim Mk.I at the Bristol Aeroplane Company's airfield at Filton, five officers and twelve Flight Masters and Wireless Operators from the Finnish Air Force arrived at Bicester on 10 January 1940 to assist in the delivery of this additional batch of aircraft. When the crews arrived at Filton to collect the aircraft they were awaiting them, complete with Finnish AF swastikas, which they hurriedly whitewashed out. Twelve Blenheims duly arrived at Bicester from Filton. An overnight downpour removed much of the whitewash and instigated a rumour in Bicester that captured German aircraft were on the airfield! Twelve crews arrived at Bicester on 21 February, to deliver the Blenheims as far as Dyce (Aberdeen), two from 21 Sqn., three each from 107 and 110 Sqns. and four from 101 Sqn. The Blenheims were delivered to their homeland by FAF crews, but were too late to make an appreciable mark on the war effort before Finland capitulated in March.

At about this time further land was acquired to the north and south of the field for construction of a concrete perimeter track. This and access tracks, two crossing the Buckingham road, totalled about six miles (10km.) and served forty-one 'frying pan' dispersals. Airfield defence was well catered for by numerous pillboxes and a defended air-raid shelter positioned at either end of each of the four hangars and four anti-aircraft sites, equipped with Lewis guns provided defence from the Luftwaffe. A Battle Instruction School was also set up in 1940. In March the Local Defence Volunteers (later renamed The Home Guard) took over the defence of the airfield.

February and March again saw periods when flying was hampered due to heavy rain and flooding. On 27 March Blenheim Mk.IV P6929 undershot on a night landing and hit the side of the bomb dump in bad weather. During that month, the Station Flight took over the task of towing targets for 6 Group squadrons from Abingdon's Station Flight and Battles were taken on charge for this task. In April, RAF policy designed to rationalise the operational training methods decreed that all squadrons in Group Training Pools were converted to Operational Training Units, the squadrons being reformed elsewhere as operational units. Thus, on 8 April, 104 and 108 Sqns. merged to become 13 OTU in 6 (Training) Group, with an establishment of 36 Blenheims and 12 Ansons. 'A' and 'B' Flights of 104

Ready to leave for Finland early in 1940 was Blenheim Mk.I BL-138
[D. S. W. Blee collection]

Blenheims Mk.I BL-138 and BL-142 for the Finnish Air Force, ready to leave Bicester early in1940.
[D. S. W. Blee collection]

Sqn became 'A' and 'B' Flights of 13 OTU and 'A' and 'B' Flights of 108 Sqn. became 'C' and 'D' Flights. On the same day Weston-on-the-Green became another satellite airfield for Bicester.

13 OTU at Bicester and 17 OTU at Upwood were to become the two major medium bomber training units. The OTU task was still to train crews to operational standards on the Blenheim for the light day bomber role, each course lasting about six months. Pilots were initially trained on the Blenheim Mk.I, many of which were fitted with full dual controls, and observers and wireless operators/air gunners on the Ansons. Later, when crews had been formed, training moved on to the Blenheim Mk.IV. In addition to the Blenheims and Ansons, three Battles, K9232, L4951 and N2023, were delivered to the unit in May for target towing duties. Sadly, the OTU's experience with the Battle was to be short lived, as K9232 suffered an undercarriage collapse at Squires Gate on 3 May, L4951 side slipped into the ground at the same location on 10 June, with the loss of three crew, and N2023 was transferred to 17 OTU at Upwood shortly after. As there were few suitable ranges for bombing and gunnery practice inland, a detachment was maintained at Squires Gate, Blackpool, for many months, from where a bombing range off the Lancashire coast was used. Later Grendon Underwood was to double as a range as well as a 'K' site. A range at North Wooton on the east shores of the Wash, had a ship shaped target, and was used for practice attacks on shipping.

May and June 1940 were blessed with very good weather, 3700 hours being flown during June, by which time an additional twelve Blenheims and four Ansons had joined the unit. His Majesty King George VI visited the station on 19 July, arriving at the railway station, where he was received by the AOC-in-C Bomber Command, Air Marshal C. F. A. Portal CB DSO MC, and the AOC 6 Group, Air Cdre. MacNeece-Foster. During his visit the King presented the DFC to Flt. Lt.Coutts-Wood, after which he left for Upper Heyford.

On 22 July the Station came under the command of the newly formed 7 Group, which had its headquarters at RAF Brampton. On 8 August, Blenheim Mk.I L6767 was badly damaged when it crashed on the field and the following day L1191 suffered loss of control and dived into the ground at Weston-on-the-Green, with the loss of the three crew members. Bicester having so far escaped the attention of the enemy, the Luftwaffe made amends on the night of 9/10 August, when a stick of sixteen high explosive bombs was dropped in a two mile line from the village of Chesterton to the Weston-on-the-Green satellite. Five of the HEs landed on the airfield, causing little damage but giving Weston the distinction of being the first Oxfordshire airfield to be bombed by the enemy.

Another accident occurred on 13 August, when L9038 suffered an engine failure on take-off and crashed, killing the sole occupant. The night of 25/26 saw a second visit from the Luftwaffe, when sixteen HE bombs and about 140 incendiaries were dropped in the area, the dummy airfield at Grendon Underwood serving its purpose by taking the brunt of the attack. Next night Weston received yet another visit, leaving the lads with another seven holes to fill in. Seven similar attacks took place between 29 August and 26 September. On 25 September Blenheim Mk.IV T1796 crashed near Lichfield when force-landing while lost. The Luftwaffe was again taking a keen interest in the area and three HE and one delayed-action bomb were dropped close to nearby Launton railway station on the same day.

On 28 September Blenheim Mk.I L6781 swung on take-off and hit trees at Weston and burnt out, the pilot escaping without injury. Sadly, the same pilot was less fortunate when he was involved in a mid-air collision with another Blenheim on 8 October during which both crews perished. The pilot of the second aircraft was Flt. Lt. Coutts-Wood DFC, who had received his decoration from the King in July.

Bicester's only war claim

On Thursday 14 October the 'Oxford Mail' reported the shooting down by British aircraft of a Ju. 88 the previous day, after it had carried out a bombing and strafing raid on a country district in the home counties. In fact the Ju. 88 had machine-gunned aircraft dispersed on the eastern edge of Bicester airfield at about 14.30 hrs and had suffered from the swift action of the AA defences. Smoke was seen pouring from the infiltrator as he headed south for the Channel and home but luck was not with him this day. He managed to limp as far as the Berkshire border and crashed on Rowded Down, to the east of RAF Harwell, some fifteen minutes later. The aircraft was Ju 88A-1 LI + LS of 8/LG and three of the four crew survived the crash. Soldiers of the Royal Artillery were justifiably able to claim the downing of the aircraft as theirs.

Weston-on-the-Green ceased to be a satellite on 1 November, and on that day the newly completed Hinton-in-the-Hedges airfield took over the role. October and November saw further visits by the Luftwaffe, which inflicted only minor damage in the area, without casualties. On 27th Blenheim Mk.IV R3705 crashed after an engine failure on take-off from Bicester and night flying began at the Hinton-in-the-Hedges satellite. The year came to an end with fog and rain greatly reducing the training programme.

1941

There was much snow and ice in January 1941, which again had an adverse effect on flying schedules. On the14th Blenheim Mk.IV N3563 crashed and burnt near Brackley and for the final ten days of the month the airfield was closed due to snow. Adverse weather conditions had soon made Hinton-in-the-Hedges unserviceable and Upper Heyford's satellite at Brackley (renamed Croughton in July) was shared for a time. On 28 February HRH the Duchess of Gloucester visited for an inspection in her capacity of Air Commandant of the WAAF. During the month Bicester was declared unsuitable for night flying, unless under ideal conditions, due to its shape and surrounding trees, but as Hinton was unserviceable, night flying was also carried out at Brackley.

March came in like a lion and went out like a lion, as the month was plagued with many crashes. On 3 March control of Blenheim Mk.IV R3751 was lost in cloud and it dived into the ground at Shillingford, Devon, while P4856 lost a propeller over the sea but made a successful landing at Squires Gate. R3813 lost a propeller on 14 March but was also fortunate in landing safely. Anson N5157 force-landed near Oxford on 19 March and four days later Blenheim Mk.IV T2285 stalled during a night overshoot at Bicester and crashed. To alleviate the problem with Hinton-in-the-Hedges, thirteen Blenheims of 'A' Flight were dispersed to Oulton in March but on 3 April moved on to Great Massingham, both of which were satellites for 2 Group in Norfolk. By 7 April, the first anniversary of the formation of 13 OTU, 26670 hours had been logged and 217 pilots, 240 observers and 273 air gunners posted out to front line squadrons.

At 23.45 hrs on 10 April a prowling Ju. 88 was shot down by night fighters, and crashed near Upper Arncott, 5 miles (8km.) south-south-east of the airfield, with the loss of all four crew. Six days later Blenheim Mk.IV V5881 flew into the Irish Sea off the Isle of Man without trace. One of the dispersed Blenheims, L1216, stalled out of cloud and dived into the ground at Little Massingham, Norfolk on 22 April and on the same night more bombs were dropped to the east of Bicester.

Blenheim fatalities due to engine problems

On 3 May Blenheim N6228 stalled on approach and crashed onto the airfield, the pilot later dying of his injuries. Two days later 'D' Flight of the OTU began operating its Anson navigation trainers from Hinton-in-the-Hedges, which was at last serviceable. A spate of unexplained accidents occurred in June, one involving Z5803, which lost its port propeller but managed to return safely to base, was similar to two incidents in March. On 20 June L9238 ditched in the Irish Sea following seizure of the port engine, the crew being rescued. Next day L9035 also lost its port propeller and force-landed near Wrexham, all three crew receiving injuries. The reason for these incidents, and possibly several others, turned out to be a faulty oil seal. Investigations revealed that a particular seal allowed oil pressure to fall below the danger level. On 23 June Blenheim Mk.IV V5751 dived into the ground from about 3000 ft near Ambrosden, killing the crew. The cause was not known but was again possibly due to engine failure. Thus ended a very black month for the OTU. July began with two VIP visits, Major Bayender, the Iranian Assistant Air Attache on 1st followed ten days later by Captain Schlegel, the Swedish Air Attache. On 15th L1203 crashed at Bicester and L6768 at Hinton-in-the-Hedges, and the following day L6767 also crashed at Bicester. The last day of the month saw L6680 colliding with a stationary Wellington after a night landing at Chipping Warden.

Construction projects were still going ahead, as on 8 May a new Officers' Mess opened on the Buckingham Road and during July the Link Trainer (Type 6414/37) became operational.

13 OTU's establishment in August was 48 Blenheims, sixteen Ansons and two Lysanders. On 17 August the pilot of Blenheim L1181 baled out after running out of fuel over Northamptonshire and on the 28th Z6099 went missing on a flight to the Isle of Man, the wreckage being found some time later on Snowdonia. An extra unit at Bicester was 7 Group Communications Flight, which moved in from Wyton at the beginning of September. The month did not pass without a fatality, as Blenheim Mk.IV V5377 dived into the ground near Frome, Somerset with loss of all crew.

Blenheim Mk.IV L9205 was possibly another victim a seal failure, when on 22 October it lost its port propeller and crashed just outside the airfield. On the same day a staff pilot ferried Blenheim Mk.I L6638 from Upper Heyford to Bicester "without having a daily inspection carried out," was unable to lower the undercarriage and was "obliged to land on the fuselage"! On 27 October a Ferry Training Flight, also known as 'X' Flight, was formed within 13 OTU to train crews to ferry Blenheims overseas, in particular to the Middle East Air Force, and desert-camouflaged Blenheims Mk.V, destined for hotter climes, now started to make an appearance at Bicester. 'A' Flight was obliged to move to Hinton-in-the-Hedges during the month to make way for the new unit. On 23 November Tutor K8171 crashed while doing aerobatics near the airfield, killing the Chief Ground Instructor and the Training Wing adjutant.

In December three crews were lost tragically when V7962 stalled on take-off with incorrect trim and crashed on the airfield on 6th, L9383 suffered an engine failure on take-off on the 10th and on the 26th P4856 spun in near the airfield. During the month, thirty-two crews passed out, all but two being allocated for Middle East units.

1942

On 21 January 1942, the Ferry Training Flight, was redesignated 1442 (Ferry Training) Flight and was to survive for another seven months before being disbanded. On 29 January Blenheim IV R3907 lost an engine and stalled out of cloud into the ground near Devizes with the loss of all three crew members, and on 6 February L6809 struck a tree after take-off from Hinton, also with the loss of the complete crew. The end of March saw two Blenheims lost without trace, Z7983 over the Irish Sea in a snow blizzard on the 28th and R3838 over the North Sea on the 31st. At the end of April 734 (Defence) Sqn. handed over the task of defending the station to 4004 Flight of the

Although seen here in 141 Sqn. markings, Blenheim Mk.I K7059 had served with 90 Sqn. at Bicester.

newly-formed RAF Regiment. The only incident for the month occurred when Blenheim Z6145 suffered a tyre burst on landing, resulting in a swing and the undercarriage collapsing. Blenheim L9206 crashed in Leicestershire with loss of all crew on 2 May and four days later L8755 was lost without trace over the North Sea.

On 11 May, 7 Group became 92 Group, the Communications Flight being redesignated accordingly, and in September it moved to Little Horwood. Forty crews completed courses during May, two going to 2 Group and 38 to 1442 Flight, which in turn sent twelve crews to the Middle East. At the end of the month the OTU was put on stand-by for air/sea rescue patrol duties over the North Sea following the thousand-bomber raid on Cologne, but in the event it was not called upon. Six of the unit's Ansons were transferred from Hinton to carry out air/sea rescue patrol duties over the North Sea on 2 June following the second thousand-bomber raid. Again on 26 and 27 June about a dozen OTU aircraft were sent out over the North Sea on patrols. June was another bad month for accidents; one Blenheim Mk.I and no less than seven Mk. IVs were

involved in force landings and belly landings at Bicester and an Anson belly landed and two more were involved in a ground collision at Hinton-in-the-Hedges. During the month 4164 hours were flown by the OTU and 308 by 1442 Flight, which received 46 more crews from the OTU, and in turn sent 51 more crews to the Middle East.

At the beginning of July, the Blind Approach Calibration Flight (BACF) moved in from Watchfield, equipped with Ansons, Masters and Oxfords for checking and calibrating the blind approach equipment now widely in use at bomber bases. That month the control tower came 'on air' when Air Traffic Control staff with communications equipment were introduced. A relief landing ground at Finmere, just over the Buckinghamshire border, had been used by 13 OTU since early 1941, but it was not until the last day of August that Finmere became a satellite, the unsuitable Hinton-in-the-Hedges having finally relinquished the task on 23 August.

August was a month of exercises, with 23 aircraft participating in a defence exercise over London on the first day of the month and eighteen aircraft in a

Blenheim L1154 of 13 OTU

similar drill two weeks later. On 28 August two Blenheims, N6169 and V6197, collided in darkness during an exercise over the North Sea. Next day Sgt. Moss, a crew member from N6169, was found drifting in his dinghy near the east coast, the only survivor from the two aircraft. The village of Stratton Audley, a little over a mile from the airfield seemed fated during September. Wellington HF865 from 22 OTU, Wellesbourne Mountford, crashed there on 7 September with the loss of five crew and on 28 September 13 OTU Blenheim Mk.IV R3761 crashed at the same location and was destroyed by fire. From September, all AA gunners and personnel allotted defence tasks within 92 Group received their weapons training at Bicester.

On the last day of October two Albermarles arrived for assessment trials with the OTU. From early October until the end of November Turweston, which had concrete runways, was also used as a satellite airfield. The Blind Approach Calibration Flight was redesignated as 1551 Beam Approach Calibration Flight on 20 November, by which time its establishment was four Ansons, two masters and three Oxfords.

In the summer of 1942 Mrs. J. D. Taylor, having completed her square-bashing, was waiting for a wireless operators' course at Blackpool, and in the meantime was posted to Bicester with some other WAAFs to spend time doing general duties. Billeted in nearby Stratton Audley, she soon became great

friends with Lilian Rolfe, who was waiting for the same course. Lilian spoke fluent French and had joined the WAAF from Rio de Janiero, where her father was working. When at Bicester she told her friend that she thought she would be of more use with her background in an intelligence capacity and had various interviews when at Bicester, but to no avail. On arrival at Blackpool she still pursued this aim, and was eventually accepted into the SOE. In 1944 Lilian was one of three female agents executed at Ravensbruck Concentration Camp for spying and although not the main character was immortalised in the film 'Carve Her Name With Pride'.

During 1942, most of the crews trained at Bicester and Upwood were destined to join Blenheim units in the Middle East, but the formation of 72 OTU in the Sudan in November took over this work. On 21 December Blenheim IV Z7361 lost its starboard engine on take-off and collided with one of the two Albermarles, P1459, destroying the latter's tail and rear fuselage. What must have been of great interest to both aircrew and groundcrew was a visit to Bicester by 1426 (Enemy Aircraft Circus) Flight during the month with an He.111, a Ju. 88, a Bf.110 and a Bf.109. Mike Henry DFC was Gunnery Leader on the OTU at the time and was able to scrounge a flight in Ju.88 HM509 on 11 December. He recalls that it circled Bicester and made dummy attacks from various angles so that the gunners could familiarise themselves with the features of the aircraft. He also remembers that the aircraft could turn within the boundary of the airfield and was certainly far more manoeuvrable than the Blenheim.

Jim Aitken, then an airframe fitter, recalls civilians from the Bristol factory carrying out modifications to Beaufighters in one of Bicester's hangars in late 1942. These could possibly have been the conversion of some aircraft to the torpedo-carrying role for 254 Sqn., which equipped with the type at the end of the year.

Blenheims Mk.V were now operating in Africa and in order to train crews for delivering replacement aircraft, 307 Ferry Training Unit (FTU) was formed at the end of December.By mid-January 1943 the unit had received six Blenheims Mk.V and ten crews had been posted in.

The middle of the war

1943 got off to a bad start when on 3 January 13 OTU Anson N5176 collided with visiting Proctor DX235 of the Air Transport Auxiliary after the brakes failed; both aircraft were later repaired. The Blenheim was now obsolescent, and the hard-pressed 2 Group squadrons were preparing for delivery of the new high-performance American- built Bostons, Havocs, Mitchells and Venturas. Initial training was to be carried out in Canada, followed by an acclimatisation period with 13 OTU. Bicester, lacking concrete runways, was unsuitable for this task and Finmere, where concrete runways were nearing completion, was to be used as an alternative.

Due to flooding at Bicester, 307 FTU carried out training at Finmere from 2 to 7 February. On 16 February Boston Mk.III Z2197 was collected from West Raynham and two days later the unit moved to Finmere. By the end of the following month the FTU had forsaken its Blenheims for two Bostons and three Havocs. In preparation for crews to deliver these new types to awaiting squadrons, an OTU crew was tasked to survey the route from Bicester to the Middle East, staging through Portreath and Gibraltar. The remaining Albermarle carried out successful experiments with night flash-photography during the month. By this time Bicester was defended by three gun positions, one with Vickers machine guns and the other two having pairs of Lewis guns.

On 15 April 1551 BAC Flt. was disbanded into 'B' and 'C' Flights of the Signals Development Unit,

Blenheims were a common sight a Bicester - more 13 OTU machines.

based Hinton-in-the-Hedges, and moved there in August and September respectively. Turweston was again used as a satellite from 1 May to 3 July. During May the unit was sporting an assortment of types at Bicester and its satellites; an Albermarle, Ansons, Blenheims Mk I, IV and V, Bostons Mk II, III and IIIA, Defiants, Havocs Mk.II, Martinets Mk I and II, Mitchells Mk.II, Mosquitos Mk.IV and VI, Oxfords and a Tiger Moth! The first operational Bostons arrived to await delivery on 12 May, and the Blenheims Mk.V, now much outdated, started to leave for Maintenance Units. The Boston's entry into service was not without problems, as one departed for North Africa on 25 April, but made a forced landing near Seville in Spain, and only three days later a 307 FTU aircraft crashed whilst on a local test flight, with loss of all crew. Mosquito conversions were also taken on by the unit in May, but lasted only a few weeks before the task was transferred to Warboys.

On 1 June, Bicester was transferred to 70 Group, Fighter Command, and 13 OTU was now earmarked to train crews for service with the Tactical Air Force. During the month, owing to lack of space in the 1937 watch office, a platform was constructed on the top floor to allow a clear view of the landing area at all times. The unit also made use of Kidlington's dome trainer for ground to air anti-aircraft gunnery practice.

Low fuel consumption on the survey flight to the Middle East had pleasingly demonstrated how economical the Boston was, and in June, further consumption tests were carried out. On 5 June Boston Mk.III W8287 flew a four-an-a-half hour sortie, recording three air miles per gallon. Another new type, the Beaufighter, joined the unit at about this time. On 30 June another ground incident occurred, when one of the based Martinets, JN417, taxied into visiting 15 OTU Tiger Moth T6312.

307 FTU at Finmere continued to train crews and deliver aircraft without incident throughout the summer, and were rewarded by a congratulatory message from AOC 70 Group, in September.

'Baron' Humphreys, a pilot who went through a course in the summer of 1943, recalls that part of it was devoted to low flying and that a final exercise was a low level flight from Dogdyke railway station, north of Boston, north-north-west across the fens to simulate bombing the Ruston-Bucyrus factory, east of Lincoln. This leg was flown as low as possible, then following the road to Newark at an altitude of 250 ft, but in reality nobody rose to that height. Each Blenheim was fitted with a camera, which photographed the altimeter every few seconds to prove the actual height. During his time at Bicester 'Baron' recalls two Blenheims returning from low-level sorties with no airspeed indication, the pitot heads beneath the fuselage being full of wheat seeds, and another returning with 150 ft of telephone wire attached to one wing. After Bicester, 'Baron' went on to Turweston and Finmere, where he converted to Bostons with crews destined for Italy. As he had only returned from the Middle East the previous December, he was instead posted to 613 Sqn. at Lasham to fly Mosquitos, and his low-flying training came in very useful when on operations.

The beginning of the end

Mosquitoes had replaced the Blenheim by January 1944, by which time 13 OTU had an establishment of nine Ansons, seven Bostons, three Martinets, 26 Mitchells and 26 Mosquitoes. In mid-January, the FTU, which had by now added Wellingtons to its strength, was absorbed into 304 FTU at Melton Mowbray. At about this time a pair of Spitfires were taken on charge for high-speed fighter affiliation work. On 1 February, 420 Repair and Salvage Unit was formed at Bicester in No 84 Group, but its stay was short lived, as it moved to Stapleford Tawney on 19 February. The final 13 OTU Blenheim flight was on 25 February and three days later the last Blenheim Mk.I departed. Since its formation in May 1940, 13 OTU had operated more than 240 individual Blenheims.

The build-up for operation 'Overlord,' the invasion of mainland Europe, was now in full swing and Bicester was playing its part. Units under the control of 84 and 85 Groups arrived on the Station and training for the Allied Expeditionary Air Force was soon under way. The Forward Equipment Unit (FEU) which was intended to be set up on the

BICESTER

Continent settled at Bicester on 17 February as part of 85 Group, 2nd Tactical Air Force (2 TAF). During April, the FEU received 2588 tons of equipment from MUs, 1420 tons of which were soon with 2 TAF units in the field. Such a large logistics operation required a lot of personnel, and thirty officers and 833 other ranks (including 190 WAAF) were on strength by the end of April. By the end of June, this had risen to 42 officers and 1035 other ranks. A Forward Transportation Section was set up at Gosport, to transfer the supplies from the FEU vehicles to shipping bound for France.

Now with a dual role, Bicester's small grass field was not suitable as the home of a busy OTU and a support unit for 2 TAF. So on 12 October 13 OTU moved over the border to Harwell in Berkshire (now Oxfordshire), which had just been vacated by the Airborne Forces units of 38 Group, which had moved to Rivenhall. Finmere remained one of the unit's satellites for the time being, however. The FEU had stockpiles of equipment around the perimeter of the airfield and in ten portable WW1 vintage Bessonneau hangars and although Bicester's days as a busy flying station were now over, Dakotas regularly visited to pick up urgent supplies. In December 1944 Bicester had a strength of 1811 RAF and 600 WAAF personnel, a total that must have been similar to the local town.

The FEU retained its name and was under the control of 12 Group (Fighter Command) until 1 January 1945, when it was retitled 246 MU in 57 Wing, 40 Group of Maintenance Command, RAF Bicester also transferring 40 Group. February saw a change in role to that of an Equipment Transit Centre. Its function now became that of a Receiving and Routing Centre, supplying the demands of 2 TAF. It now comprised a holding unit for aero-engines, a transit centre for equipment for the continent, a motor transport (MT) assembly point and a servicing centre for replacement MT and radio vehicles. To give an idea of the size of the Bicester operation, during the week ending 27 January, 936 tons of equipment were issued. On 1 March, a Mechanical Transport Servicing Unit (MTSU) was formed at Bicester, and in early August a detachment of 40 Group HQ moved to the Station.

In July 2nd TAF became British Air Forces of Occupation, and the task of supplying units in Germany ceased in September. Bicester then became a Receiving, Storage and Issuing Depot (RSID) for MT vehicles. On 30 November 1946 a sub-site of 246 MU was set-up at Hinton-in-the-Hedges.

Early post-war life at Bicester

In January 1947, Leon Prisley, a sergeant engine fitter, was posted to 246 MU and on arrival during a heavy snowfall, was most surprised to find that on an airfield covered with lines and lines of MT vehicles of all descriptions there were no aircraft. He had been posted in to help to prepare these vehicles for service with other units, at home and overseas. His next surprise was finding that half the workforce consisted of ex-aircrew who knew nothing about motor transport and West Indians who were all awaiting demobilisation and were devastated by the bitter winter. In February 40 (Maintenance) Group moved in from Andover and in May, the Station became a sub-site of 3 MU, an Aircraft Equipment Depot, based at Milton near Didcot. The sub-site at Hinton-in-the-Hedges held a sale of some 1015 vehicles during the year which realised £168,000, and on 30 November, closed its gates for the last time, all 246 MU responsibility returning to Bicester. Changes during 1949 saw 246 MU disbanded on 1 April and by the end of the year the Transport Command Parachute Servicing Unit was in residence.

A new unit, the Civilian Supply and Technical Officers' School (CSTOS), was formed within the station establishment on 3 September 1951, becoming a self-contained unit on 1 April 1952. An American Ammunition Depot, 282 MU Explosives Storage Depot, was formed at Bicester on 18 August 1952 with sub-sites at Finmere and Wing, but before the end of the year it had moved to Finmere.

A new role as a Repair & Salvage Unit

71 MU, a Repair and Salvage unit within 43 Group, which had been disbanded at Little Horwood in November 1947, was reformed at Bicester (as a lodger unit on a 43 Group station !) on 15

Flown to Bicester from the Middle East after service with 233 Sqn., Valett VW821 was then employed to demonstrate the use of airbags to lift aircraft. [B. Martin]

December 1953. The unit had been given a few weeks to settle in and establish itself, not being required to become fully operational until 1 February 1954. Its responsibility was for the salvage and repair of aircraft in much of southern England, including the counties of Bedfordshire, Berkshire, Buckinghamshire, Essex, Hertfordshire, Kent, Middlesex, Oxfordshire, Surrey, Sussex and the County of London. At this time the three services had many times the number of aircraft that they do today, and consequently accidents were more numerous. Additionally, 71 MU was called upon to house and maintain aircraft preserved for public exhibition, a task previously carried out by 58 MU at Honington. On its reformation, 71 MU was to have an establishment of eleven officers and 218 other ranks, but by the end of December only four officers and 71 other ranks had arrived. By January 1954 this had risen to seven officers and 160 airmen and the unit had by now made eight salvage tasks. Teams had been sent to three different locations to carry out Cat. 3 aircraft repairs, fifteen transportation tasks had been carried out and the MT section had recorded 21338 miles. The Exhibition Flight was now up and running and in February it had a Venom, four Vampires, a Meteor, a Bleriot and three Derwent and two Avon engines on charge. The AOC 43 Group, Air Cdre. F. G. S. Mitchell CBE, visited the Station on 2 March, and formally presented the MU with its unit badge at a parade outside the hangars. Of note with the MU in July were Wellington T.10 6592M and Oxford HN420 and in November a

Bicester's gate guardian in 1958 was Spitfire LF.16E TE356, carrying 34 Sqn. codes 8Q:Z. This aircraft became 6700M in March 1967.

Varsity T.1 WL633 [U] of 2 ANS was written off at Manston on 23 January 1958 and the remains were brought to Bicester before eventual sale to a scrap merchant.

Mosquito T.3 of the Ferry Training Unit which had force landed at Benson the previous month and Venom FB.l WK437, also from Benson. The CSTOS had departed by the end of the year.

1954 had proved to be a hectic learning period for the MU, and by January 1955, after the first full year of operations, of some 548 tasks placed on the unit, only 26 had still to be completed. The MT Section had recorded 300311 miles. In February the new AOC of 43 Group, Air Cdre. Moore CBE, visited the Station. On 4 June, Sunderland MR 5 RN288 of 201 Sqn had bounced on landing in swell and nosed in off of Eastbourne, Sussex following a fly-past during the RAFA Annual Conference in the town. 71 MU was tasked with clearing the wreckage, during which, an ACl was seriously injured, and subsequently lost a leg.

The gliders arrive

71 MU's parentage transferred from 43 Group on its disbandment, to 41 Group on 2 January 1956, and 40 Group, with HQ at Bicester, was absorbed by 42 Group on the same day. January also saw the formation of 'The Windrushers Soaring and Gliding Club' on the Station, the first flight of a club glider taking place on 1 February. This was the start of a long association between the gliding fraternity and RAF Bicester, which is still thriving today. Control of the inactive satellite of Finmere was transferred from the USAF to Bicester on 6 March. In August, The Channel Islands, Hampshire, The Isle of Wight and Northants were added to the already large area of responsibility of 71 MU. To give the locals an insight into what was going on over the fence, the Station held its first Public Open Day on 11 August. The weather was not perfect but 2000 plus visitors saw gliding displays and a supply drop by parachute and had the opportunity to sample the delights of gliding themselves. On the ground were static aircraft and exhibitions by Oxford University Air Sqn., Transport Command Parachute Servicing Flight, the Mountain Rescue service and others, a

He.162A 120235 at Bicester on 29 October 1961. It is now in the IWM as '6' of JG.77

Javelin FAW.1 XA549 [M] awaits its fate at Bicester on 29 October 1961 after service with 87 Sqn. In February 1962 it was given M serial 7717M for use as a ground instructional airframe.

parade by the Bicester Hounds and sporting events including football and polo.

In August 1957, 71 MU again expanded its coverage, when Cambridgeshire, Huntingdonshire, Leicestershire, Norfolk, Rutland and Suffolk came under its umbrella. From 31 July, the inactive station at Roade, Northants., came under control of Bicester until disposed of in December of the following year. The size and responsibility of the unit justified the rank of the CO being raised from a Sqn. Ldr. to a Wg. Cdr. in October.

Bicester ceased to be a sub-site for 3 MU, Milton in May 1958 and in June, WE600, the Auster C.4

which had taken part in the 1957/58 Trans-Antarctic Expedition, joined the exhibition aircraft held on the station. On 23 July, 6204 Bomb Disposal Unit took up residence, when it moved in from Stafford. In keeping with many units at this time having 'gate guards,' by now Spitfire LF.16e TE356 was displayed on the parade ground, where it remained until the late sixties.

After many years without any powered flying, Bicester came alive again on 12 January 1959, when Oxford University Air Squadron (OUAS) Chipmunks arrived from Kidlington, the unit's HQ going to Little Rissington.

Camouflaged Gnat F.1 XK740, previously operated by the Ministry of Supply, at Bicester on 29 October 1961

Me.163B 191660 is comprised of major components of several of its type
and is now in the IWM as Yellow 3 of 1/JG400

The 1960s

The area of responsibility for 71 MU was again redefined in 1960 as that area of the United Kingdom south of a line drawn by the southern boundaries of the counties of Merioneth, Montgomeryshire, Shropshire, Staffordshire, Leicestershire, Rutland and Lincolnshire. On 15 August a Durham UAS cadet was killed when his Chipmunk crashed at Milcombe, south of Banbury, whilst undertaking a solo flight from Bicester to carry out level flying, spinning and stalling exercises.

January 1961 saw the arrival of yet another non-flying unit, when the Maintenance Command Group Defence School arrived from Calshot. Wellington T.10 MF628, which is now in the RAF Museum, arrived for painting in January and departed in March. A report in June noted an abundance of Spitfires, K9942, P2617, PK274, TE536, in storage, and TB287 and TB308 in the scrap compound. Gliders noted at the time in the RAFGSA hangar were Olympia 2Bs, T.21 Sedburghs, two Kranich, Grunau Babys an EoN Baby, a Skylark and an EoN 460.

During July, notice was given that on 1 October, RAF Bicester would disband and that 71 MU would become the Commanding Unit of the airfield. On 28 July, 40 Group HQ disbanded, and control of the Station came under HQ Maintenance Command at

Apparently in top condition, Hunter F.1 7422M was previously WT684 and had been grounded in 1957 after use by 229 OCU at Chivenor.

RAF Andover. At 1500 hrs on 15 August, a first took place, when Jet Provost T.3 XN600 became the first (and possibly only) jet-powered aircraft to land on Bicester's grass airfield. It had flown in from 27 MU, RAF Shawbury, for 71 MU to dismantle and re-erect at the Battle of Britain Exhibition at Richmond Terrace, London. On 19 September, the Under Secretary of State for Air, the Right Honourable Julian Amery MP, in company with the Air Secretary, Air Chief Marshal Sir Theodore N. McEvoy KCB CBE, visited the Station to see the workings of the MU.

The RAFGSA moves in

71 MU duly took over the running of the Station on 1 October, and with it the responsibility for administrative and technical facilities for OUAS, Transport Command Parachute Servicing Flight, 6204 Bomb Disposal Flight, Maintenance Command Ground Defence School, the off-Station RAF Liaison party at RAF Upper Heyford and the RAF Careers and Information Centre at Oxford. The month concluded with the 'Windrushers' Gliding Club holding a successful public open day on the 29th. At around this time the RAFGSA HQ was set up as a lodger unit.

An assessment of the value of the use of helicopters in assisting salvage teams was held at Bicester in April 1962. Results were favourable, some heavy and awkward loads being lifted, but whether or not they were used in the field is not known. April was a busy month as the RAFGSA held a practice meeting from the 14th to the 23rd, during which Bicester was host to some 130 personnel, 25 gliders and eight Chipmunk tugs. A lot was learnt from this meeting, in preparation for the British National Championships, which were held at RAF Aston Down in June.

1963 started with bad weather throughout January and most of February. The OUAS Chipmunks could not fly during January, but a ski equipped Beaver and a helicopter did brave the elements to visit. On 28 February a ceremony was held in the Officer's Mess to mark the long association between the town of Bicester and RAF Bicester. During the pleasantries, the Station Commander was presented with a plaque bearing the Arms of Bicester from the Chairman of the Urban District Council and he in turn presented the Chairman with a 71 MU shield, to be hung in the council chamber. In June, the OUAS departed to RAF Leuchars for annual summer camp and in turn the following month they were to play host to Southampton UAS. During October, the RAFGSA Centre was integrated as part of 71 MU and the first of 169 new married quarters was handed over for occupation. During the year the RAFGSA acquired the first of its many civil registered glider tugs, a couple of Auster 6As, which were converted to Tugmasters.

A gliding competition was held by the RAFGSA during Easter week, 1964, and in April the RAFGSA spent a fortnight at Nympsfield. In May a party was sent to Lasham for the National Gliding Championships, the CFI, Flt. Sgt. Andy Gough

With engine cover open for inspection, Chipmunk T.10 WP 919 [A] of Oxford UAS awaits attention at Bicester in 1966. This Chipmunk was sold in September 1975

After towing targets around Maltese skies for several years, Beaufighter TT.10 RD867 was sent to Bicester and is seen here being refurbished in May 1965. [B. Martin]

taking 5th place.

In April 1965 the RAFGSA again held an Easter Meeting, the weather unfortunately reducing flying to only four days of the ten allocated. A Naval pilot crashed at Weston-on-the-Green, and luckily only sustained minor injuries. HM The Queen annually made a formal visit to one of the RAF Commands, and this year it was the turn of Maintenance Command, with Bicester the selected unit. This visit took place on 10 May, when the Queen and HRH Prince Phillip inspected the unit. The Royal Party arrived at 12.50 to be greeted with a Royal Salute, given by the Guard of Honour from the Queen's Colour Squadron of the RAF Regiment. The RAF Central Band was also in attendance, and played throughout the ceremony. This was the Station's third Royal visit.

A visit to the unit on 6 May, the rehearsal day, noted OUAS Chipmunks WP919 [A], WK638 [B], WZ860 [C], WP785 [D], WD359 [E], and WK575 [F] and assorted gliders, including Blaniks 274 and 326 lined up on the grass. Demonstrating lifting methods was Wessex HC2 XS678 with underslung Chipmunk WD382, Javelin XA701 [X] under lifting

tackle, Valetta C1 VW821 of 52 Sqn. with airbags beneath the wings and Valetta C.1 VW188 loaded on a Queen Mary to demonstrate the method of transport. Outside one of the hangars were two Bloodhounds and inside was Beaufighter TT.10 RD867, recently returned from Malta, Pembroke C.1 XF798, Spitfire X4590, Hurricane P2617, Mosquito TV959, Gnat T.1 XP530 Javelin XH837, the nose of Canberra WH903, an anonymous Lightning, Vampire T.1 1 and another Gnat. The inmates of another hangar were Lysander JR:M, Sopwith Camel F1921, Gladiator K8042, Spitfire K9942, Bristol Boxkite replica and Hunter XF946. The RAFGSA hangar revealed Austers (with roundels) G-ARGI, G-ASEF and G-ASOC, Chipmunks WP803 [G], WP964/Z and WB560, a Super Cub and rare based Miles M.18 G-AHKY. The wreck of Spitfire Mk.XIV RM694 was noted on the dump.

The OUAS ventured south, to St Mawgan, for that year's summer camp in June and the following month Cambridge UAS visited Bicester. In August the RAFGSA was busy, when it hosted the Junior Inter-Services Championships, and held an ab-initio and an instructors course. 4,192 movements were

Demonstrating what a Wessex HC.2 could do, XS678 is seen lifting Chipmunk WD382 at Bicester on 6 May 1965. [B. Martin]

recorded in August, 59 military, 76 transit, 2 civil, 1214 aero tows and 2841 glider winch launches. Bad weather curtailed flying for many days during the final three months of the year and throughout January 1966. During the month a Royal Navy Wessex was forced to land at Bicester due to bad weather when on its way from Lee-on-Solent to Linton-on-Ouse, and was not able to depart for two days.

The Inter-Service Gliding Championships were held at Bicester during 1966 Easter week, when appalling weather conditions allowed flying on only one day. During April the RAFGSA CFI, Flt. Sgt. Andy Gough, was awarded the BEM for 'Outstanding Services to RAF Gliding'. On 23 June OUAS departed to RAF Marham for summer camp and in July the Wales UAS arrived for a four week camp. This year's Inter-Service Gliding Championships took place from 22 to 29 August.

The Transport Command Parachute Servicing Unit, which had been in residence since 1949, departed for RAF Hullavington in January 1967, making way for 5 Light Anti-Aircraft Wing, RAF Regiment, which formed from 1 (LAA) Sqn. and 2 (LAA) Sqn., which had arrived recently from RAF Butterworth, Malaya, and RAF Changi, Singapore, respectively. The UAS summer camp was at Woodvale, Edinburgh UAS Chipmunks taking their place in the Bicester circuit. Tragedy shocked the station on 4 July when a Bocian glider crashed on the airfield shortly after take-off, killing the front occupant and seriously injuring the rear seat occupant. This accident did not portray gliding in a good light in local papers as only three months earlier a T. 21 had crashed at nearby Weston-on-the-Green, killing both occupants.

On 27 September Merlyn Rees, Secretary of State for Defence (RAF), visited the unit, but before

Sqn. Ldr. H. B. Iles kept Miles M.18 Series 2 G–AHKY at Bicester. Here it is seen carrying racing number 39. 'KY had been impressed into RAF service as HM545 and was sold to the then Flt. Lt. Iles in March 1948.

The RAFGSA's Auster 6A G–ASOL, complete with roundel in December 1966.

he arrived at the Station he was shown around the new married quarters which were under construction in the town of Bicester. Towards the end of the year the Exhibition and Transport Flights began to prepare many historic aircraft for static display at the following year's RAF Golden Jubilee celebrations.

During November the Salvage and Transportation Flight recovered the wreckage of Iberian Airways Caravelle EC-BDD from its crash site in Sussex and transported it to Farnborough for investigation. It had crashed on 4 November, during descent to Heathrow, with the loss of 37 lives. On 10 November a Wales UAS Chipmunk force-landed near Weston-on-the-Green and was recovered by a Wessex helicopter to Bicester. The Maintenance Command Ground Defence School departed in November, by which time RAF Bicester had been operational for fifty years. On 1 December another RAF Regiment squadron, 26, arrived from RAF Changi. With the withdrawal of the Beverley from RAF service in late 1967 seven of these popular workhorses were flown into Bicester for breaking up. During 1969 most of the Beverleys disappeared from the Bicester skyline when they were broken up by a local scrap dealer, although XH122 was to survive until 1971, in use for air portability training with the Army as 8045M.

Wrecks and relics

Early in 1970 5 (LAA) Wing, RAF Regiment disbanded. In April the RAFGSA added two former

The six Chipmunks of Oxford UAS at Bicester on 6 May 1965 — WP919 [A], WK638 [B], WZ860 [C], WP785 [D], WD359 [E], and WK575 [F].
After service elsewhere, WP785 was sold as G–BBWI and WD359 as G–BBMN. [B. Martin]

Beverley C.1 XM103 was one of several of its type stored at Bicester after withdrawal from service. This one had been used by 84 Sqn. in the Middle East.

crop spraying Chipmunks for use as glider tugs and later in the year, two Falkes, their first powered gliders. The other two RAF Regiment squadrons departed to Germany during the year, 26 to RAF Gutersloh on 1 July and 1 to RAF Laarbruch on 1 September. 1971 saw the unit busy with the largest restoration project carried out at Bicester, when Lancaster 1 R5868 was being prepared for static display at the RAF Museum.

A familiar site alongside the Buckingham road for a couple of years after flying in on 25 October 1972 was the former Boscombe Down-based Hastings C.1 TG500, which was used to give salvage crews experience in lifting crashed aircraft with pneumatic air bags.

1974 was to be a year of massive defence cuts, the White Paper published on 19 March sealing the fate of Bicester as the home of 71 MU. Together with 60 MU from Dishforth, 71 MU was to transfer to Abingdon and be merged into the Engineering Wing there. During the year the RAF disposed of many of its aging Chipmunks and the RAFGSA took the opportunity to add more of the breed to their tug fleet, which not only operated from Bicester but also at other service gliding clubs which came under the RAFGSA umbrella. Another Chipmunk, G-AOSU, which had undergone a partial conversion with a Lycoming power plant, was also acquired and the

Withdrawn from use by 84 Sqn., Beverley C.1 XL130 [Y] awaits its fate at Bicester in 1968.

BICESTER

Hastings C.1 TG500 on arrival at Bicester from A&AEE, Boscombe Down,
on 25 October 1972. Notice the nose radar 'pimple'.

conversion completed at Bicester. Its first flight was made in April 1975, the Lycoming 0-360 engine proving a much more economical proposition than the original Gipsy Major. A 100 hour test programme was then carried out and in April 1976 the Association received the approval of the CAA for the conversion and all Chipmunks were subsequently converted.

During 71 MU's time at Bicester, there was always a flourishing scrap yard on the airfield, with the remains of broken-up and crashed aircraft. Before the move to Abingdon, in 1975 there was a massive clearance of the yard and by the end of the year only Jet Provost XN633, Hastings TG500 and Whirlwind XK937 remained. With the run-down of Bicester, OUAS was also to move to Abingdon and on 31 January, a farewell flypast was made by four Chipmunks. On 31 March 71 MU eventually made

the move to Abingdon and RAF Bicester closed down, leaving only the RAFGSA on the site. After a few weeks of Care & Maintenance under Abingdon's control, the Station was handed over to the Army on 20 May, becoming a barracks.

In 1977, when HRH The Prince of Wales presented the 1976 British Gliding Team with the Prince of Wales Cup at the Royal Aero Club's annual prize giving, he expressed an interest in doing some glider flying. Arrangements made, he duly made his first glider flight at Bicester on 8 June 1978. Weather prevented the planned 100 km triangular flight but he was able to do four aerotows, accumulating about two hours local flying in a Twin Astir. On 22 November, the unit again became RAF Bicester, for use by the United States Air Forces in Europe (USAFE) as a storage unit.

In anticipation of an escalating Gulf War the

Slingsby Cadet TX.3 basic glider XN243 at Bicester on 15 September 1996.

The 'Fort' type watch office (to drawing 1959/34) at Bicester was never superceded by a more modern building, remaining in use for all air traffic control purposes.

USAFE had converted some of the buildings into offices and a medical storage facility and turned the domestic site into an Emergency Military Hospital, complete with hospital wards, operating theatres and a burns unit. Fortunately they were not required. With the closure of Upper Heyford in 1994, the USAFE medical facility at Bicester also closed. The Cherwell District Council commissioned a survey to assess the extent and quality of the surviving buildings and structures on the unit, and this was published in August 1996. As a result many of the buildings are protected by preservation orders.

RAF Bicester in 1999
During May 1 999 a Vintage Flying Week for gliders was held, leading up to and including Saturday 29

Above: The contents of a Bicester hangar on 15 September 1996, with Ask.18 R48 in the foreground.
Below: Chipmunk 22 G–ATVF, with a Lycoming engine, at Bicester on 29 May 1999.

Ask.21 sailplane R21 of the RAFGSA on 15 September 1996.

May, when the RAFGSA held its 50th Anniversary celebrations. The week produced a varied assortment of vintage types, and on the Saturday Robs Lamplough's Spitfire, P-51 and Sea Fury put in an appearance, although the Sea Fury only made a circuit of the field. As part of the 50th Anniversary celebrations an aerobatic team was formed to take part in the RIAT display at RAF Fairford in July. A dual aerobatic sequence was performed by ASK 21s R25 and R73 followed by a solo demonstration by an Lo-100 high performance glider, the RAFGSA at last having opportunity to perform for the aviation world and general public. What was believed to have been the last Inter Services Regional Gliding Competition held at Bicester took place from 10 to 19 August.

Early in 1999 it was rumoured that RAF Bicester was to close by the end of the year. Leading up to the eventual decision, the RAFGSA had been tentatively planning purpose- built accommodation on the northern edge of the airfield. As the year progressed it was speculated that Henlow would be the probable base for the RAFGSA, but eventually it was announced that Little Rissington would be the site for the relocation. At the moment it is understood that accommodation will not be

Above: The C-type hangar used by the RAFGSA in July 1997.

Right: An A-type 'aircraft shed' at Bicester in July 1997, after fencing had been erected.

The guardroom and fire party building (to drawing 959/25) at Bicester in boarded-up condition in September 1996.

completed until late in 2000.

Today the RAFGSA Centre at Bicester, under the auspices of the Centres manager, retired Sqn. Ldr. Ted Norman, has three areas of responsibility – the everyday running of the RAFGSA, providing a depot for the storage and maintenance of RAFGSA gliders and tugs and the running of the Joint Services Adventure Training Gliding Centre. The RAFGSA provides gliding facilities at a number of locations for those in the three services. At Bicester this is mainly at weekends, but on weekdays the Joint Services Adventure Training Centre provides tri-service training courses, places being allocated in line with the strengths of the three services - 50% from the Army, 15% from the Navy and 35% from the RAF. In 1996 RAFGSA membership totalled 921. A co-located unit and associated member of the RAFGSA is the Oxford University Gliding Club, which has its own gliders, but students can fly RAFGSA gliders when under instruction. Apart from the C-type hangar used by the RAFGSA, most of the other buildings on the site are now fenced-in.

USAF personnel from RAF Croughton still occupy some of Bicester's remaining married quarters and the Defence Clothing and Textiles Agency (DCTA) have occupied premises on the domestic site since the mid-nineties. The current plan is that the airfield will be sold by the MoD in March 2000 and Wantage-based Grove 2000, whose chairman is Robs Lamplough, has made an offer to

An aerial view of RAF Bicester, taken on 4 September 1999 by the author. This shows nicely the conventional layout of a 1930s Expansion Scheme airfield, which at Bicester was never much altered.

purchase the site for redevelopment. Proposals made by Grove 2000 include the purchase of the whole site and refurbishment of existing buildings and letting them for business use; the creation of a huge recreation facility to include eight football pitches or similar, indoor tennis courts and rifle ranges, and an indoor go-carting track. A circular walk and an aviation heritage centre, which would include gliding and would retain at least part of the site as an active airfield. Although these plans would provide the town of Bicester and the surrounding area with a much-needed recreation facility, they have not been favourably received by Cherwell District Council, which has earmarked the site for 1,860 houses to be built by the year 2011. The MoD has so far rejected the Grove 2000 offer but discussions are still continuing with Cherwell District Council. Bicester airfield is still with us after more than eighty years and will at least remain active until the 21st century.

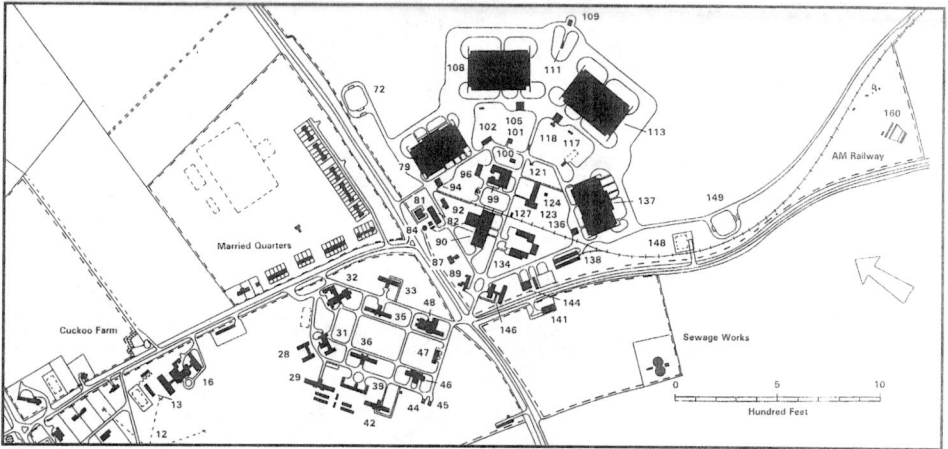

BICESTER
Site Plan 1939

12 - W/T block	81 - Reservoir	118 - Petrol Tanker shed
13 - Qtrs block for single officers	82 - Power-house	121 - Fire Tender & Huck starter
16 - Officers Mess and quarters	84 - High level water tank	shelter
28 - Qtrs Block for Sgt pilots	87 - Fire Party House	123 - Station Armoury
29 - 'E' Type barrack block	89 - Guardroom & Fire House	124 - Shelter for Pracice Bombs
31 - Sergeants Mess	90 - Main Stores & Church	127 - Lubricanting oil installation
32 - Institute	92 - Parachute Store	134 - MT Sheds
33 - Single storey barrack block	94 - Petrol Tanker Shed	136 - Petrol Tanker Shed
35 - 'E' Type barrack block	96 - Lubricant Store	137 - 'A' Type aircraft shed
36 - 'E' Type barrack block	99 - Main Workshops	138 - Coal Yard
39 - Airmans garages	100 - Technical latrine	141 - Pyrotechnic Store
42 - 'E' Type barrack block	101 - Link Trainer	144 - Works Services Building
44 - Mortuary	102 - Engine Test House	146 - Station Offices
45 - Ambualnce garage	105 - Petrol Tanker Shed	& Operations Block
46 - Station Sick Quarters	108 - 'C' Type aircraft shed	148 - Aviation petrol installation
47 - Ration Store	109 - Watch Office & Tower	149 - Aviation fuel installation
48 - Dining Room	111 - Fire Tender shelter	160 - Machine Gun Range
72 - Aviation Fuel Installation	113 - 'C' Type aircraft shed	
79 - 'A' Type aeroplane shed	117 - Pyrotechnic store	

OFFICIAL NAME - Bicester   LOCAL NAME - Bicester

COUNTY:	Oxfordshire	**OBSTACLES:**	nil
LOCATION:	1.75 mls NNE of	**RUNWAYS:**	SE/NW 3600 ft. grass
	Bicester/		NE/SW 3300 ft. grass
LANDMARKS:	Brickworks 6 mls E	**HOUSING:**	Permanent
GRID REF:	SP598245	**HANGARS:**	two A-type,
LAT:	51° 54' 53" N		two B-type
LONG:	01° 08' 00" W	**OPENED :**	1917
CONTROL TOWER:	1959/34	**CLOSED:**	(still open for gliding
HEIGHT ASL:	270ft		in 1999)
LIGHTING:	nil	**CURRENT USE:**	-
AIRFIELD CODE:	BC		

UNITS PRESENT AT BICESTER

UNIT	CODE	FROM	DATE IN	DATE OUT	TO	AIRCRAFT USED
118 Sqn.		Netheravon	7.8.18	7.9.18	(disbanded)	(various)
44 Trg. Depot Stn.		Port Meadow	1.10.18	8.19	(redes. 44 Trg. Sch.)	Avro 504; F2B; Pup
2 Sqn.		Genech	2.2.19	9.19	Weston-on-the-Green	FK.8
44 Trg. Sqn.		(ex 44 TDS)	8.19	12.19	(disbanded)	??
5 Sqn.		Hangelaar	8.9.19	20.1.20	(disbanded)	nil
100 Sqn.		Spittlegate	10.1.28	14.4.28	Weston Zoyland	Horsley
100 Sqn.		Weston Zoyland	15.5.28	3.11.30	Donibristle	Horsley
33 Sqn.		Eastchurch	5.11.30	27.11.34	Upper Heyford	Hart
101 Sqn.		Andover	1.12.34	6.5.39	West Raynham	Sidestrand; Overstrand; Blenheim
48 Sqn.		(formed ex 101)	25.11.35	16.12.35	Manston	
144 Sqn.		(re-formed)	11.1.37	9.2.37	Hemswell	Overstrand; Audax
90 Sqn.	TW	(formed ex 101)	15.3.37	10.5.39	West Raynham	Blenheim
217 Sqn.		Tangmere	16.8.37	11.9.37	Tangmere	Anson
12 Sqn.	PH	Andover	9.5.39	2.9.39	Berry-au-Bac	Battle
142 Sqn.	KB	Andover	9.5.39	2.9.39	Berry-au-Bac	Battle
108 Sqn.	LD	Bassingbourn	17.9.39	8.4.40	(disbanded to 13 OTU)	Blenheim; Anson
104 Sqn.	EP	Bassingbourn	18.9.39	8.4.40	(disbanded to 13 OTU)	Blenheim; Anson
13 OTU	**	(formed ex 104 & 108 Sqns.)	8.4.40	12.10.44	Harwell	Battle; Martinet; Blenheim; Havoc; Anson; Lysander; Albemarle; Beaufighter; Spitfire; Tiger Moth; Defiant
7 Gp. Comm. Flt.		Wyton	9.41	11.5.42	(redes. 92 Gp. CF)	Proctor
1442 (FT) Flt.		(formed ex 13 OTU Ferry Flt.)	21.1.42	1.8.42	(disbanded)	Blenheim
92 Gp. Comm. Flt.		(ex 7Gp.Comm.Flt.)	11.5.42	14.9.42	Little Horwood	(various)
Blind App. Cal. Flt.		Watchfield	3.7.42	20.11.42	(became 1551 Beam App. Cal. Flt.)	Anson; Oxford
1551 BA Cal. Flt.		(ex BA Cal. Flt.)	20.11.42	15.4.43	(disbanded into SDU)	Anson; Master; Oxford
307 Ferry Trg. Un.		(formed)	24.12.42	18.2.43	Finmere	Blenheim
Sigs. Dev. Unit ('B' Flt.)		(ex 1551 BA Cal.Flt)	15.4.43	28.8.43	Hinton-in-the-Hedges	Anson
Oxford UAS		Kidlington	12.1.59	26.9.75	Abingdon	Chipmunk
RAFGSA Centre			1.10.63	(current)		

** AT, EV, FV, KQ, OY, SL, UR, WO, XD and XJ